GLENEAGLES
SCHOOL LIBRARY

WIND POWER

Published by Smart Apple Media
1980 Lookout Drive
North Mankato, Minnesota 56003

Design and Production by EvansDay Design

Photographs: Richard Cummins, JLM Visuals
(Richard Jacobs, Breck Kent), Sally Myers, Tom
Myers, Jim Steinberg

Copyright © 2002 Smart Apple Media.
International copyrights reserved in all countries.
No part of this book may be reproduced in any
form without written permission from the publisher.

LIBRARY OF CONGRESS CATALOGING-IN-PUBLICATION DATA
Gibson, Diane, 1966–
Wind power / by Diane Gibson.
p. cm. — (Sources of energy)
Includes index.
Summary: Describes wind energy and how it is
generated and used. Includes a simple experiment.
ISBN 1-887068-79-1
1. Wind power—Juvenile literature. [1. Wind power.]
I. Title. II. Series.
TJ820 .G53   2000
621.4'5—dc21   99-055895

FIRST EDITION
9 8 7 6 5 4 3 2 1

SOURCES OF ENERGY

windpower

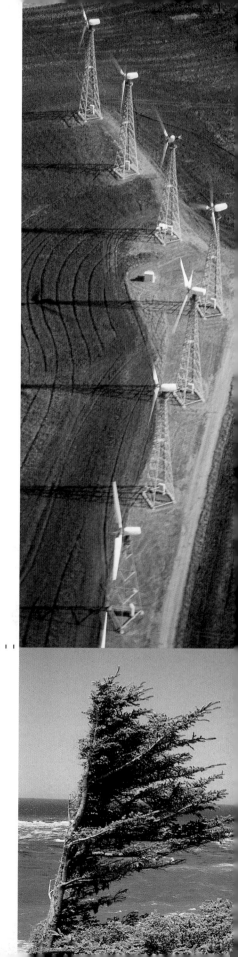

DIANE GIBSON

windpower

All over the world, the wind blows. It races over the land, making flags flutter and puffy clouds float across the sky. It moves warm air to cold places, and it brings coolness to hot areas. People can use the wind as an energy source. It can push boats forward, draw water up from the ground, or generate electricity. The power of the wind can be harnessed in many ways.

WHY THE WIND BLOWS

THE WIND BLOWS because of the sun. As it shines, the sun warms the earth's **atmosphere**. The sun does not heat the whole planet evenly, however. Some places get warm more quickly than others. Heated air weighs less than cold air, so warm air rises. Since cold air is heavier, it sinks. As the warmer air moves up, the cooler air rushes in to take its place. The movement of the cooler air is what we know as wind. You can observe this movement of air for yourself by opening the door of your refrigerator just a crack. Down near the floor, you can feel the cold air blowing out. This is similar to the way the wind blows on the earth.

WHEN THE AIR FROM WARM AND COLD CLIMATES MEETS AND MINGLES, THE RESULT IS WIND.

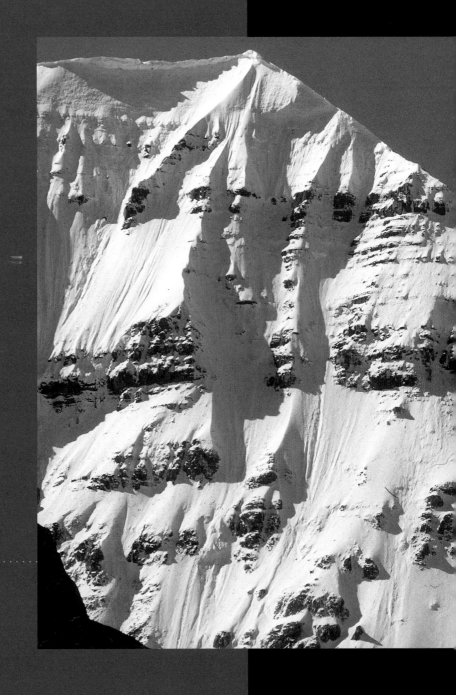

The force of constant winds can give shape to landscapes and even living trees.

At the South Pole, the wind can blow up to 200 miles (322 km) per hour. On April 12, 1934, the wind was measured at 225 miles (362 km) per hour at Mount Washington in New Hampshire.

WINDMILLS

Windmills are machines that are powered by the wind's energy. Most of them look like giant pinwheels. Windmills have blades, called vanes, mounted in a circle. The vanes catch the force of the wind, making the windmill spin. For centuries, windmills have been used to pull water up from the ground. At the top of a windmill, the vanes are connected to gears. The gears turn a large shaft, or pole, that runs all the way to the bottom of the tower. At the end of the shaft is a giant screw that is buried in the ground. As the windmill spins around, it turns the gears, which turn the shaft. This in turn spins the screw. Ridges around the screw pull the water up from the ground. Windmills are still used today. In fact, about one million of these pumps can be found on farms around the world. The water brought up from the ground is used to **irrigate** crops or to provide drinking water for people and animals.

Wind farms may be found anywhere the land is open and the wind blows steadily.

Almost one-third of all the world's wind-powered electricity is produced in California. Windmills were once used to cut wood. A large saw moved up and down as the wind turned the mill's vanes. Workers cut the wood by carefully pushing it against the saw.

One of the tallest wind turbines is on Magdalen Island in Canada. It stands 361 feet (110 m) high and creates enough electricity to light 40,000 light bulbs at the same time.

Despite their different forms, all wind turbines include vanes that rotate a central shaft.

MAKING ELECTRICITY

⊙ Wind can also be harnessed to create **electricity**. Machines that do this are called wind turbines. They are similar to windmills, but there are a few differences. Many turbines have fewer vanes (usually two or three) that spin around. The vanes are made of metal or plastic and are slightly curved to catch the wind better. Not all wind turbines have vanes that stick out like arms. One type looks more like an eggbeater than a pinwheel. ⊚ As their vanes spin around, the turbines power other machines called generators, which create electricity. Inside a generator is a large, round **magnet** that spins around on a pole connected to the turbine. Around the magnet are thousands of coils of wire. Electrical power is produced as the magnet spins around inside the coils of wire.

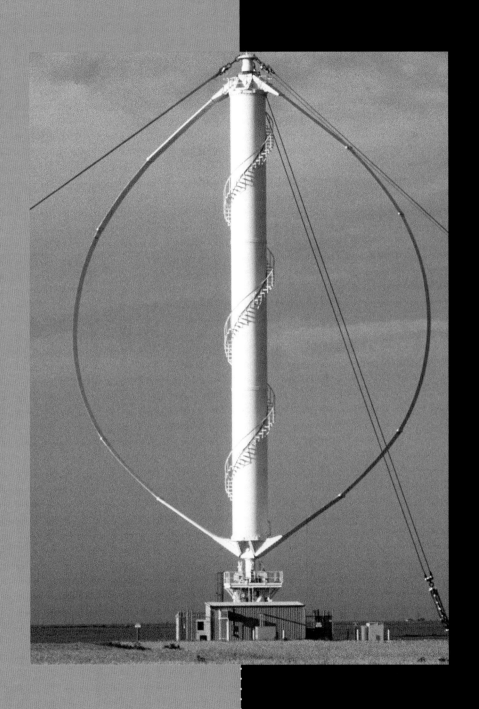

THIS LARGE, EGGBEATER-SHAPED TURBINE GENERATES ELECTRICITY AS THE WIND SPINS ITS VANES.

WINDMILLS HAVE LONG BEEN USED TO DRAW WATER FROM THE GROUND, OFTEN FOR STORAGE IN TANKS.

The nations of Denmark and Germany both use wind farms to create some of their electricity. At one time, Holland used thousands of windmills. The windmills pumped seawater out of the land to keep it from flooding.

WIND FARMS

MANY WIND TURBINES set up together to make a lot of electricity are known as a wind farm. These farms are set up in places where the wind blows all the time. California, Hawaii, and New Hampshire are some of the best states for wind farms. Computers control the turbines on these farms. The computers determine the direction the wind is blowing, then turn the tops of the turbines to face into the wind. If the wind blows too hard, the computers turn the turbines off so they don't break. The electricity created by wind farms can be used immediately. It may be sent to homes and businesses through cables and wires, or it may be stored in **batteries** for later use. The biggest wind farm in the world is located in Altamont Pass, California. This farm can produce 637 **kilowatts** of power an hour. Most of the power is purchased and distributed by a large utility company.

Before the invention of engines, sailors relied on the wind to power their ships.

During the 1800s, some large ships had as many as 25 sails to catch the wind. **There are about 20,000 wind turbines in the world. Almost 15,000 of them are in California.**

MORE WIND-GENERATED ELECTRICITY IS PRODUCED IN CALIFORNIA THAN ANYWHERE ELSE IN THE WORLD.

THE FUTURE OF WIND POWER

Wind power is not yet widely used. There are several reasons for this. First, the wind doesn't always blow when we need it to. Second, wind farms can be expensive to build and maintain. They take up a lot of space, so they are often built in places that are far away from big cities. Also, the generators on wind turbines make a lot of noise. Wind farms that have hundreds of wind turbines can get very loud. Still, using the wind's energy is a good idea. Today, we rely heavily on fuels such as coal and oil, which must be burned to produce energy. Burning these fuels pollutes the air. Also, the earth has only a limited supply of these fuels; one day we will run out of them. Wind-powered machines, on the other hand, do not pollute the air. The more they are used in place of burning fuels, the cleaner our air will be. And as long as the sun shines to warm our planet, we will always have the wind.

Although wind farms can be quite noisy, they provide a very clean source of energy.

Many recreational boaters have fun on the water by raising a sail instead of starting a motor.

Some boats are equipped with both sails and engines. This allows sailors to cut down on the amount of fuel they burn.

HISTORY

One early use of wind was to power boats. People sewed animal skins together to make a large sheet called a sail, then mounted it on a boat. As the wind blew, it filled the sail and pushed the boat forward. Sails are still used on many boats today. Old drawings have been found in Egypt that date back to about 5,000 years ago. The drawings show sailboats being pushed down the Nile River by the wind. The 15th-century explorer Christopher Columbus traveled across the Atlantic Ocean on a huge type of sailboat. Large sailing ships of his time also had oars so sailors could row the ship if the wind stopped blowing.

People have relied on wind power for thousands of years and will continue to do so in the future.

EXPERIMENT

⊙ **Watch Heat Rise** This experiment will let you see warm air rising. You will need:

A plastic bottle
A bowl
A balloon
Ice-cold water
Hot water

⊙ Fill the bottle with ice-cold water. Allow the water to sit in the bottle for at least a minute.

⊙ Fill the bowl with hot water. (Ask for an adult's help in handling the hot water.)

⊙ Now empty the cold water from the bottle and secure the balloon over the bottle opening. Carefully set the bottle in the bowl of hot water and watch the balloon fill up with air.

⊙ The cold water in the bottle causes the air inside to become cold, too. But when the bottle is placed in the hot water, the air is warmed and moves upward. Since it can't escape, it fills the balloon.

GLOSSARY

An **atmosphere** is the air and gases that surround a planet.

Batteries are containers filled with chemicals that can store or produce electrical power.

Electricity is a type of energy used in our homes to run lights and appliances.

Farmers **irrigate** crops by supplying them with water brought in through pipes and channels.

A **kilowatt** is a measurement of electrical power. One kilowatt equals 1,000 watts.

A **magnet** is a piece of metal that attracts iron and steel.

INDEX

A
atmosphere, 6

B
batteries, 15

C
California, 10, 15, 17
Columbus, Christopher, 21
computers, 15

E
electricity, 10, 11, 12, 14, 15

G
generators, 12, 18

I
irrigation, 9

N
New Hampshire, 8, 15

P
pollution, 18

S
sailboats, 16, 17, 20, 21
sun, 6

W
wind farms, 10, 14, 15, 18, 19
windmills, 9, 10, 14
    cutting wood, 10
    drawing water, 9, 14
    preventing floods, 14
wind power
    history of, 21
    problems of, 18
wind speeds, 8
wind turbines, 11, 12, 13, 15, 17, 18